*A seventeenth-century brick bee bole shelters a straw beehive or skep in the 'Secret Garden', a feature of the recreated Tudor garden at the Tudor House Museum, Southampton. To the l-f-* ...... 'er holds two more skeps.*

T0174474

# BEE BOLES AND BEE HOUSES

A. M. Foster

Shire Publications Ltd

# CONTENTS

Published by Shire Publications Ltd,
PO Box 883, Oxford, OX1 9PL, UK
PO Box 3985, New York, NY 10185-3985, USA
Email: shire@shirebooks.co.uk   www.shirebooks.co.uk
© 1988 by A M Foster.
First published 1988.
Transferred to digital print on demand 2014.
Shire Library 204.    ISBN: 978 0 85263 903 0.
Typeset in Times
Printed and bound in Great Britain.

British Library Cataloguing in Publication Data available.

ACKNOWLEDGEMENTS
The author wishes the thank all those who have helped in the preparation of this book, especially Dr Eva Crane, M. J. C. Smith, D. R. Parker, and those who have kindly provided photographs or allowed photographs to be taken. These include the National Trust, National Trust for Scotland, Mr G. Hawthorne and the Berkshire College of Agriculture, Mr Bernard Mobus of the Northern College of Agriculture, the Gloucestershire College of Agriculture, the Tudor House Museum, Southampton, the Somerset Rural Life Museum, E. H. Thorne (Beehives) Ltd, the Hereford City Library, and the many private property owners who gave us the freedom of their homes and gardens. Without the assistance of our field workers, there would be many fewer illustrations and special thanks are due to Phil and Stan Mercer, Graham Mair, Malcolm and Janis Mair, Paul Campbell, Jim MacKenzie, May Miller, Anne Newell, Richard Whittington, and Caroline and Jeff Whiteway for their help. Illustrations are acknowledged as follows: Hereford City Library, pages 3, 8 (top right), 9, 20, 21(top); Bernard Mobus, page 14 (bottom); Somerset Rural Life Museum, page 6; E. H. Thorne (Beehives) Ltd, pages 2, 7 (top). All other illustrations are by the author.

COVER: *Photograph courtesy of Giles Booth.*

BELOW: *In this nineteenth-century drawing of 'Mr Blow's Bee Garden' a great variety of beehives and bee houses is illustrated including two straw skeps at the lower right of the picture. Behind the fence at the right are two bee houses; on the front of each is a row of entrances for the bees.*

*A beekeeper extracts honey from frames in this early twentieth-century photograph from the Alfred Watkins Collection in the Hereford City Library. In the background is a bee house. Watkins, a local beekeeper, recorded many examples of Herefordshire beekeeping methods before and during the 1920s.*

# INTRODUCTION

'My son, eat thou honey, because it is good,' exhorts Proverbs 24:13, and for centuries mankind has prized this product of the honeybee for its sweetness and its culinary, preservative and medicinal benefits. There is written evidence that the Romans kept bees in Britain; bees had probably been domesticated for many centuries before Roman times. Written records also exist of beekeeping in the Anglo-Saxon and Norman eras. From the early medieval period a different kind of evidence appears in the form of structures created by generations of beekeepers to house and protect their hives. Where these structures were made of wood few now remain, but where they were of brick and stone they are still to be found, sometimes in a ruinous state but

very often in as good condition as when they were first used, perhaps centuries ago.

Beekeeping was not the prerogative of any one section of society and these remains are found not only in the gardens of great houses but also in the outbuildings and walls of farms and small cottages. Some structures appear in ones and twos, others in large groups; some are simple, others fanciful. Over nine hundred have been recorded in the United Kingdom and it is certain that many more are still to be found. This book is a brief guide to what is known and the author hopes it will encourage readers to look about them and perhaps to find yet more relics of the beekeeping past.

3

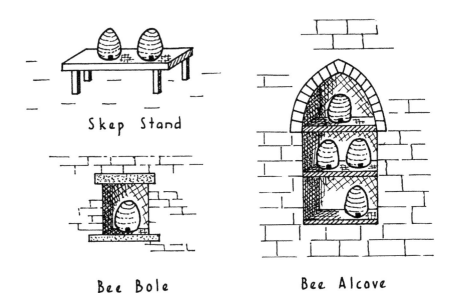

Skep Stand

Bee Bole

Bee Alcove

Bee Shelters

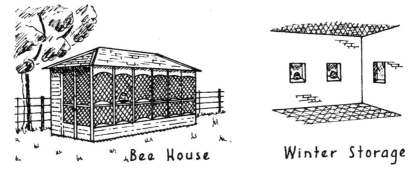

Bee House

Winter Storage

*Structures for sheltering skeps.*

# BEEKEEPING PAST AND PRESENT

The basis of modern beekeeping is the movable-frame hive. This consists of a large, usually wooden, bottomless box, the brood box, in which the bees raise their young and upon which are placed one or more shallow boxes, the supers, for the storage of honey. Between the brood chamber and the supers is a perforated partition called the queen excluder which allows passage of the smaller worker bees but prevents the larger queen from entering and laying eggs in the honey chamber. The whole sits on a floorboard which has one side open providing an entrance for the bees into the brood chamber. This floor may project beyond the brood box to form an alighting board upon which the bees land before entering the hive. The hive also has a roof.

Within both the brood box and the supers are suspended wooden frames filled with sheets of beeswax embossed with the hexagonal honeycomb pattern. The bees use these sheets as a foundation upon which to construct the cells in which eggs are laid and honey is stored. The frames are separated from each other and from the sides of the box by the so-called bee-space, ¼ to ⅜ inch (6-10 mm) wide. If the space were smaller than the exact bee-space it would prevent the bees' movement throughout the hive; if it were larger, additional comb would be built across the space affixing the frames to each other and to the sides of the boxes, making their removal difficult.

It was not until the mid nineteenth century that the significance of the bee-space was utilised to develop the movable-frame hive. Previously, beekeepers in Britain had used the basket hive or skep. The earliest examples were probably of wicker and were mostly taller than wide and roughly conical in shape. They were superseded by skeps of coiled straw whose tight construction provided an element of weatherproofing not present in wicker skeps. The earliest known remains of such a straw skep were uncovered at the Coppergate excavations in York and were dated to the twelfth century.

Straw skeps were much smaller than modern hives. Eighteenth- and nineteenth-century apiarists generally advocated skeps 12 inches (30 cm) in diameter and 9-12 inches (23-30 cm) high. Beekeepers wanted small hives because they depended upon swarming, the result of overcrowding in hives, to replenish their stock of bees. An eke or ring of coiled straw could be added underneath

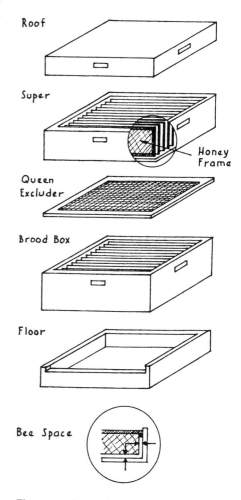

Roof

Super

Honey Frame

Queen Excluder

Brood Box

Floor

Bee Space

*The parts of a modern movable-frame hive.*

5

the skep so that the bees could expand honeycomb production into it. A cap could be placed on top for the same purpose. Either could be removed without disturbing the main hive.

Unlike modern hives, skeps were not provided with frames and foundation wax. The bees, therefore, attached comb to the inside walls of the skep and the only way to collect the honey was to remove the bees and cut the comb away from the skep. The most commonly used methods of removal resulted in the death of the colony either by drowning or suffocation. The skeps would then be cleaned out and filled with swarms the following spring. Swarming occurred when the parent hive became over-crowded, inducing the queen bee and some of the workers to fly out in search of new quarters, leaving a daughter queen to carry on the old hive. Swarms could be captured and introduced into new skeps, where they would establish themselves. The bees could be ejected from their old hive into an empty one without killing them by driving. In this procedure an empty skep was suspended at an angle above a full one and the bees encouraged to leave the lower by pounding on the walls of the comb-filled skep.

It was costly and wasteful to kill bees in order to obtain the honey and in the seventeenth century attempts were made

*An engraving from 'Modern Beekeeping' (1880) showing the capture of a swarm. The woman beating on a dustpan is 'tanging the bees', a procedure thought to calm them.*

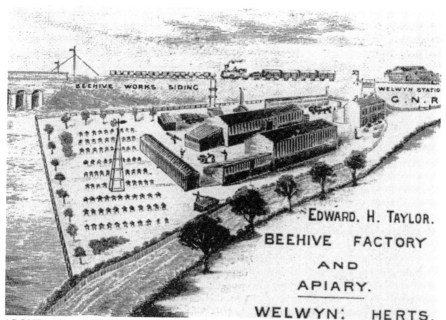

EDWARD. H. TAYLOR.

BEEHIVE FACTORY

AND

APIARY.

WELWYN: HERTS.

ABOVE: *The activity in this engraving of a nineteenth-century beehive factory illustrates the importance of honey production.*
RIGHT: *A Nutt collateral hive is protected by the verandah of a house in this drawing from J. C. Loudon's 'The Suburban Gardener' (1838). In such a hive, the central box was for brood, the adjoining ones for honey.*

to overcome the problem by providing separate wooden honey chambers either on top or to the side of wooden brood chambers. The former were known as tiered hives, the latter as collateral hives. Tiered hives, such as the Stewarton, used thin wooden strips or top-bars fixed across the top of each box under the cover and the bees built their comb downward from these bars. However, there was no allowance made for a bee-space along the sides of the box and so the bees still attached comb there. It was not until 1851 that the Reverend L. L. Langstroth in the United States replaced the top-bars with a rectangular frame which hung downwards but remained clear of the walls of the hive by a bee-space. Such frames could easily be removed. In 1862 the new hive was introduced into Britain and it became the basis for the modern hives seen today.

ABOVE LEFT: *A bee canopy shades a collateral hive in this drawing from J. C. Loudon's 'The Suburban Gardener'.*
ABOVE: *A simple stone skep stand supports a handmade wooden hive in a photograph from the Alfred Watkins Collection.*

LEFT: *A bee stone on its pedestal. The projecting tongue acts as an alighting board for the bees to land upon before entering the hive.*

*A group of skeps, each with its protective straw hackle, sits on a raised platform with rails to prevent the skeps overturning. From the Alfred Watkins Collection.*

# SKEP BASES AND THEIR SUPPORTS

Most skeps were placed out in the open on stands and so had to have some protection from wet weather. Wicker skeps were cloomed — covered with a mixture of clay and cow dung or plain cow dung which dried and hardened to shed the rain. Straw skeps were protected with a variety of covers from straw hackles to old cream-separating pans and even cabbage leaves! An elaborate method of covering the hive may be termed a 'bee canopy' and was described by J. C. Loudon as a 'roof open on every side, the props being rustic pillars and the roof being covered with thatch, reeds, woodmen's chips, spray, bark, heath or similar materials'.

For protection from damp ground, special benches and stools were constructed for the skeps to sit on and placed in a sheltered part of a garden or farmyard. Most were of wood and, whilst the majority of these have not survived, nineteenth-century photographs show them to have ranged from very simple wooden platforms and stone slabs sup-

ported on rough-hewn wooden legs to carefully constructed beds with low parapets to prevent the skeps being accidentally dislodged. Not all apiarists approved of placing skeps close together on benches. John Keys wrote in *The Antient Bee-Master's Farewell* (1796) that it was wrong to do so, as such practices were 'always the source of mistakes, quarrels, and often slaughter' as the bees attempted to enter the wrong hives. Bee stones, circular stone stands with a projecting tongue which acted as an alighting board for bees, have survived. Three examples, still in place on a Wiltshire farm, are supported by pyramidal stone pillars 23 inches (58 cm) high. The bee stones themselves are 24 inches (60 cm) across at their widest point. Their attractiveness confirms the eighteenth-century view of A. Pettigrew that 'a row of well-thatched bee-hives, all nicely clipped, standing in a cottage garden conveys to the minds of people passing by the idea of comfort and profit'.

*Nineteenth-century brick bee boles near Netherton, Oxfordshire. Two hold empty skeps. The hole at the rear of each bole may have been for ventilation when the front was covered for overwintering.*

*Twelve Welsh bee boles in a dry stone wall. Each recess is approximately 22 inches (56 cm) square and would have held a single skep.*

*Stone bee boles on a Welsh farm. The projecting edges provide additional protection from wet weather.*

# BEE BOLES

By far the most common survivor of those structures designed to house skeps is the bee bole. 'Bole' is in origin a Scots word meaning a recess in a wall. A bee bole was an integral part of the wall, open only at the front, in which a straw skep was placed to shield it from wind and rain. Bee boles are also known as bee holes, bee shells (Cumbria), bee keps (Cumbria), bee niches (Derbyshire), bee walls (a Gloucestershire term which refers to a wall in which there were a number of boles), bee houses (Yorkshire), bee boxes (Kent) and bee garths (garths also refers to the enclosure in which the boles were situated). Bee boles were not the usual method of housing skeps and Samuel Bagster commented in *The Management of Bees* (1834) that the most inefficient way of keeping bees was to 'procure a ledge in a wall'. However, the architect J. C. Loudon in 1822 listed among the recommended amenities for a labourer's cottage 'a nitch in the wall of the south-east front of the house, to hold two or more beehives...'

Between 1953 and 1983 the International Bee Research Association recorded over eight hundred sites in Britain where bee boles were still found (although no longer used). Their profusion in areas such as Tayside, Cumbria and Kent may reflect attempts at beekeeping which could only be successful in such wet or windy places if extra protection were given to the skeps.

Bee boles were constructed most frequently in orchard and garden walls but are also found in the outside walls of houses. Their close proximity to the house, whilst at first surprising, reflected the advice of seventeenth- and eighteenth-century apiarists: 'Let the hives be set as near the dwelling-place as conveniently can be, or to rooms most occupied, for the reddier discovery of rising swarms, or to be apprized of accidents. Besides, the bees habituated to the sight of the family will become less ferocious and more tractable.'

Equally important was the need to ensure that the skeps with their valuable

LEFT: *Four bee boles in a house chimney, about 1580. There are also three recesses in the cellar for winter storage of skeps.*
ABOVE: *A detail of the bee boles illustrated on the left.*

honey were not stolen. Honey could be used to pay rents and tithes and in some parts of Britain such 'honey rents' were quite common. At Mickleton, County Durham, there are late eighteenth-century records of the bee bole owner's ancestors paying tithes with bee swarms. Honey fairs were a feature of some areas and it is recorded that in the thirteenth century Edward I bought a cask of honey at Conway fair. This fair was still well known in the nineteenth century.

Also, especially before the Reformation, beeswax was valued for its use in making candles for the great churches: 'The origin of bees is from paradise,' said the Gwentian Code, 'and because of the sin of man they came thence; and God conferred His grace upon them, and therefore the mass cannot be sung without the wax.'

An interesting feature, seen especially in Scotland, is the provision of small holes in the wall immediately flanking the boles for the attachment of a bar to secure the skep within the bole and prevent theft. The bars were also used to hold in packing around the skep when overwintering the hives (see chapter on winter storage).

The walls enclosing bee boles were usually constructed of the most readily available building material, whether stone, brick or cob. The shape of a bole was governed by the materials used, those in stone and cob being mainly square or rectangular and those in brick often gabled or arched. The size of the recess was also determined by the building material. In general, bee boles were 18-30 inches (46-76 cm) in height, 15-26 inches (38-66 cm) wide, and 14-21 inches (36-53 cm) deep. Those built in brick walls tended to be smaller than the average because the walls were generally thinner than stone walls. Instances of boles as shallow as 9 inches (23 cm) have been recorded; however, shaped stone bases were sometimes inserted in the recess under the skep. These bases projected out beyond the wall and acted as alighting boards for the bees as well as supporting the skep. The majority of bee boles still to be seen have flat bases. The backs of boles are also usually flat although rounded backs occur in some stone examples.

Most recesses are found in sets, with five to six boles per property. Some sets consisted of many more than six recesses;

ABOVE: *An early nineteenth-century slate 'beehive wall' with six bee boles in a field near Saltash, Cornwall.*

RIGHT: *A bee bole under a stairway at Pitmedden House, Grampian (National Trust for Scotland).*
BELOW: *A close-up of the Pitmedden bee bole. Note the stone skep base still in the bole; such bases were used, most often in very shallow boles, to support the hive.*

13

ABOVE: *Part of the garden wall containing thirty bee boles at the seventeenth-century manor of Packwood House, Warwickshire (National Trust).*

LEFT: *Before the Reformation, beeswax exceeded honey in value and many bee boles, such as this one at Pluscarden Abbey, Grampian, were associated with religious establishments. Beeswax was needed for church candles, some of which weighed as much as 30 pounds (14 kg).*

the Tudor garden at Thornbury Castle, Avon, had 26 bee boles, now filled in. Packwood House, Warwickshire, still has thirty boles lining a south-facing garden wall. The usual distance between recesses was approximately 20 inches (50 cm) but this could vary, especially when the boles were seen as a design feature of a garden as at Thornbury, where they are set a uniform 3 feet (90 cm) apart. Often boles in sets were not placed on the same level but in tiers, the lowest tier being 2-3 feet (60-90 cm) off the ground. This was a convenient height for the beekeeper to inspect his hives. Surprisingly, some are set high under the eaves of houses. Although difficult for the beekeeper to reach, such boles would make the flight line of the bees well above the comings

14

*A thatched chalk wall near Broad Hinton, Wiltshire, with recesses for twenty skeps.*

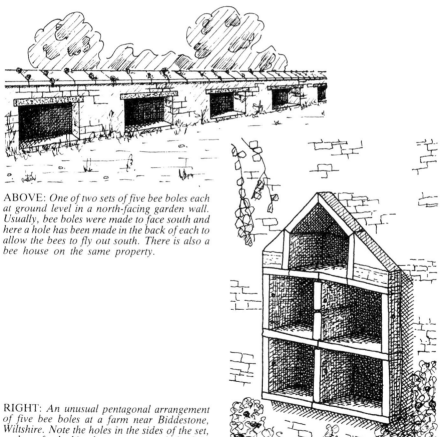

ABOVE: *One of two sets of five bee boles each at ground level in a north-facing garden wall. Usually, bee boles were made to face south and here a hole has been made in the back of each to allow the bees to fly out south. There is also a bee house on the same property.*

RIGHT: *An unusual pentagonal arrangement of five bee boles at a farm near Biddestone, Wiltshire. Note the holes in the sides of the set, perhaps for locking bars to prevent theft of the skeps.*

ABOVE: *A set of eight projecting bee boles in brick. Bricks have been laid on edge to form a gabled arch. In contrast, the smaller boles in Kent have simple gables formed by two bricks laid lengthwise.*
LEFT: *A pillar at the junction of two garden walls holds three bee boles one above the other. Many sets of boles are found arranged in this fashion.*

and goings of the inhabitants, and the overhanging roof would provide additional protection for the skeps. Rarely, bee boles also occur at ground level.

Bee boles tend to be found in south-facing walls. This orientation provided warmth for the hive and also encouraged the bees to begin work early, as recorded by the Roman writer Columella: 'Bees' dwelling-places...ought to be so arranged as to face the south-east in order that the bees may enjoy the sun when they go out in the morning and may be more wide-awake; for cold begets sloth.' Often, special care was taken to provide plants such as lavender which were attractive to bees. Sir Thomas More grew rosemary in the garden of his Chelsea house and 'lette it run all over my walls because my bees love it'.

The dating of bee boles is often diffi-cult because of the problem of determining the age of the walls in which they are built. House and garden walls were not always built at the same time and it is apparent from the use of different materials within the same wall that some have been repaired or rebuilt. Where bee boles have been filled in, their original outlines can usually still be seen. The earliest bee boles are probably those found in Tudor brick walls such as those at St Stephen's, Canterbury, Kent. The heyday of bee bole construction was, however, the eighteenth and nineteenth centuries.

Bee boles appear on all kinds of properties from the simplest cottages to the great estates. In Beatrix Potter's *The Tale of Jemima Puddle-Duck* (1908) one of three bee boles at her Cumbrian farm is illustrated. The variety of boles reflects the wealth of individual property owners.

Those associated with more well-to-do cottages and manor houses are often finished with well shaped lintels and arches of stone. Each of eight bee boles in a sandstone wall at Coton, Staffordshire, has a small decorative carving above. Moreover, there is evidence of local fashion in bee boles, as in north-west Wiltshire where at least two sets are composed of five boles each, arranged in a pentagon. The whole set projects as a unit some 3 inches (8 cm) from the wall and is surmounted by a gabled roof.

Today, while bee boles no longer hold skeps, many still contribute to the charm of a garden, filled by their owners with pot plants or small statues. This has also helped to preserve many of these attractive recesses to the present day.

RIGHT: *A modern use for a sixteenth-century bee bole in a walled garden.*
BELOW: *These eight bee boles in a garden setting illustrate their decorative as well as their functional aspect.*

*These recesses in a walled garden in St Andrews, Fife, illustrate the difference in size between a bee bole, which holds a single skep, and a bee alcove, which can hold more than one. The decorative shields, blue and white, may have been a device of the Priors of St Andrews and the whole structure was possibly made of materials taken from nearby St Andrews Cathedral.*

# BEE ALCOVES

A variation of the bee bole may be called the bee alcove, a larger, shelved recess for holding more than one skep. By 1983 41 such alcoves were known at twenty sites in the United Kingdom. Almost half of these sites have more than one alcove. At Aldeburgh, Suffolk, four alcoves together could have accommodated as many as sixteen skeps. Originally there were three more alcoves, now demolished.

Most alcoves were larger than a bole, being generally 60 inches (1.5 metres) high, 50 inches (1.3 metres) wide and 21 inches (53 cm) deep. Some are at ground level, while others range from 9 to 45 inches (0.23-1.1 metres) above the ground. The average is 24 inches (60 cm). The sides of alcoves often projected slightly from the wall into which they were built and their tops were often arched.

Like bee boles, alcoves were made to face south and, with the exception of one site, all those known are in garden walls. At Hurst, North Yorkshire, a set of eight alcoves was built into a hillside in a field. Alcoves appear in the same areas as bee boles, those where rain and wind are a feature of the climate.

Unlike boles, alcoves are a feature almost exclusively of the houses of the gentry. An elegant specimen recorded at Daresbury, Cheshire, has two adjacent alcoves fronted by Doric pillars. The stone facade above is decorated with a skep and bees carved in relief. An unusual feature at Mickleton, County Durham, was the provision of mushroom-shaped stools upon which the skeps sat within the alcove.

ABOVE: *A very simple domed eighteenth-century alcove in a stone wall near Haddington, Lothian. The wall is thickened at the back to accommodate the alcove and the remains of a modern shelf are visible near its base.*

ABOVE RIGHT: *This handsome stone Scottish alcove decorated with a classical pediment could have housed four skeps. The modern shutters replace earlier ones, which would have been closed in winter or very wet weather to protect the skeps.*

RIGHT: *An elegant eighteenth-century alcove with a carved facade depicting a skep and bees. It is at Daresbury, Cheshire.*

19

*A stone bee shelter in Hereford, photographed in the early twentieth century by Alfred Watkins. Several straw skeps and a wooden hive stand on the shelves.*

# BEE SHELTERS

Throughout Europe, probably the most common device for sheltering skeps was the bee shelter. It was a roofed, usually lean-to structure without individual niches for the hives, which were placed side by side on shelves of wood, stone or slate. Confusingly, bee shelters were also sometimes known as bee houses (Cumbria and Essex), bee butts (Cumbria), bee holes (Yorkshire) and hive shelters (Gloucestershire). *The Country House-wife's Garden,* published in 1618, describes such a shelter as 'a frame standing on posts with one floor (if you would have it hold more Hives, two floores) boarded, laid on bearers and back posts, covered over with boards, slatwise. And although your Hives stand within a hand-breadth of one another, yet will the Bees know their home. In this Frame may your Bees stand dry and warm.'

Although only about sixty examples were known by 1983 many more have probably collapsed or been removed, having been built, in the main, of wood.

The majority of those that survive are of stone. As these shelters, unlike bee boles, could easily be built on to an existing wall, it is likely that their use was much more widespread than the sixty examples imply.

An early bee shelter has survived from the sixteenth century but most were built in the eighteenth and nineteenth centuries. Their use was not restricted to any social class; they are found in the gardens of cottages, farms and large country estates. The average shelter could hold up to six skeps but those in Cumbria were generally smaller, holding three to four skeps. The large free-standing stone shelter now at the Gloucestershire College of Agriculture could hold 28 skeps.

The Cumbrian shelters are very simple and at Troutbeck, Windermere, is a typical example. The shelter, facing south-east, measures 43 inches (1.1 metres) high at the front, 50 inches (1.3 metres) high at the back, 40 inches (1 metre) wide and 25 inches (64 cm) deep.

Inside, a slate shelf could have held up to three skeps. A more substantial structure near Cockermouth, Cumbria, was free-standing with a hinged wooden door at the front and back. There were flight entrances in the front door. The provision of doors made it possible to over-winter skeps with the addition of packing, as was done in some Scottish bee boles. In addition to a door, a shelter at Auchterhouse, Tayside, also had an iron locking bar across the front, another characteristic of many Scottish boles.

Some shelters were partially recessed into the wall against which they stood and protruded from the wall face while others extended beyond the back of the wall. One, at Sykeside, Grasmere, Cumbria, extends beyond both wall faces as though a gap were deliberately left in which to place the shelter.

Most shelters, although sturdily made, are not very decorative, but a wooden

RIGHT: *A wooden bee shelter has been placed high up under the projecting roof tiles of this Hereford house. Note the decorative woodwork at the bottom edges of the shelter. From the Alfred Watkins Collection.*
BELOW: *A typical Cumbrian bee shelter at Troutbeck. Built against a house wall, it probably held three skeps on its slate shelf.*

one at Holmwood, Surrey, built in the seventeenth century, bore wooden carvings, including a depiction of the farmstead on which it stood. The most elaborate shelter is that now at the Gloucestershire College of Agriculture, Hartpury. It formerly stood at Hive House, Nailsworth, Gloucestershire, and was built before 1500 of Caen stone from Normandy, which had an ecclesiastical link with the manor of Minchinhampton of which Nailsworth was part. It is very extensively carved and has two rows of fourteen niches for skeps. A further five or even ten skeps could have stood in recesses at the base.

*The superb pre-1500 stone bee shelter at the Gloucestershire College of Agriculture, Hartpury, now housing a variety of hives.*

*The wooden bee house at Attingham Park, Shropshire (National Trust). Skeps can be seen behind four of the semicircular openings.*

# BEE HOUSES

'Bee house' is the name given to a free-standing building within which bee-hives were kept and which, unlike free-standing bee shelters, permitted the beekeeper to move around inside the building for servicing and inspection of the hives. A bee house had the advantage of allowing a number of hives to be placed within a relatively restricted space. Most bee houses could hold thirteen hives, many more than alcoves, and more than the usual number found in a set of bee boles. The bee house at Appleby Hall, Humberside, built around 1830, accommodated 32 hives, a practice of which John Keys would not have approved! A modern example built in 1947 for queen rearing could house up to one hundred small colonies.

Most extant bee houses appear to have been built in the nineteenth century. Wooden hives as well as skeps have been kept in them but bee houses were originally made for skeps and at Appleby Hall

remains of skeps were found in the bee house in 1953.

The skeps stood on shelves or individual skep stands within the bee house. The bees entered through regularly placed slits or holes in the outside wall of the house, then through the individual hive entrances behind. Often, alighting platforms were placed on the outside wall of the bee house under the entrances.

Since few cottagers could afford or indeed have needed so many hives, bee houses are in general found in the precincts of large estates and grand houses. At some sites there are bee boles as well. At Kitlochside, East Kilbride, Strathclyde, eight hives could have been kept within the bee house and a further two within bee boles nearby. Bee houses are distributed throughout Britain but their rarity, with only twenty-three known up to 1983, indicates the cost of their building and maintenance.

Most bee houses were of timber, ex-

LEFT: *This rendered stone bee house at Kitlochside, East Kilbride, Strathclyde, could have held up to eight hives, each supported on two stone pillars inside. It is the only circular example yet discovered in Britain.*
RIGHT: *A close-up of one of the eight flight entrances in the Kitlochside bee house.*

cept in Scotland where all known examples are of brick or stone. While the majority are rectangular, some have six, eight or even ten sides. The stone bee house at Kitlochside is circular, measuring 11 feet (3.4 metres) in diameter, with walls 8 feet (2.4 metres) high. Its full height, from the ground to the weather-cock at its summit, is some 12 feet (3.7 metres).

'Bee-houses may always be rendered agreeable and often ornamental objects,' wrote J. C. Loudon. The ornate design of many is an indication of the wealth of their owners. The eighteenth-century bee house at Bretforton Manor, Evesham

*This brick Sussex bee house may be contemporary with the adjoining farmhouse built in the early seventeenth century. It could have held twelve skeps, each with an entrance and small wooden alighting board (enlarged).*

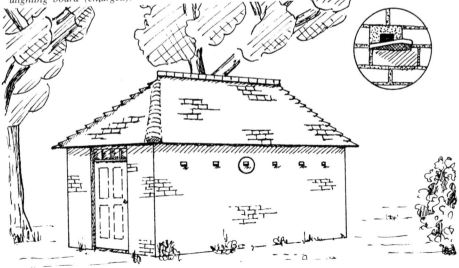

24

*The bee house at Bretforton Manor, near Evesham (Hereford and Worcester). The 'leaded' (actually iron) windows once held glass and small wooden alighting boards can be seen to either side of each window. Under the stone tiles the roof is thatched.*

(Hereford and Worcester), for example, contrasts elegant diamond-shaped leaded windows with a deliberately rusticated facade decorated with a Y-shaped pattern of split saplings. It is of brick and wood and the roof is thatch which has been overlaid with stone tiles. Inside is a bench for skeps or wooden hives and the bees entered the house through small holes on either side of each of the six windows. Each entry has in front of it a semicircular wooden alighting board. Placed in a setting of trees a short distance from the manor house, it must, in its prime, have considerably enhanced the estate grounds. There are now plans to renovate it.

At Hall Place, Burchett's Green (now the Berkshire College of Agriculture), the ten-sided bee house set in a wide open space and flanked with trees could easily be mistaken for a Victorian summerhouse with its curved roof of zinc-sheathed wood topped by a pointed finial and wooden latticework on the sides. Below each of the nine windows equipped with wooden shutters is a flight hole for the bees. The bee house is built around a great timber post only roughly shaped from its original tree trunk and the laid floor is enlivened with coloured fragments of brick and stone. It is obvious that many owners regarded their bee house as an ornamental addition to the garden, perhaps surpassing its practical use as a method of keeping bees.

The decline of estates and the money necessary to maintain such fragile buildings as bee houses has contributed to the unfortunate disappearance of many. In Hereford and Worcester, four of the five recorded are now known only from plans or photographs.

*The rebated edges of these bee boles at Pilmuir House, Lothian, were used for inserting a door to shelter the skeps during winter.*

# WINTER STORAGE

In the autumn beekeepers emptied the lightest and the heaviest skeps, the former because the bees would not have had enough honey to overwinter, the latter because they had produced the most honey and therefore gave the greatest profit. Hives of middling weight would be kept but needed extra protection from winter cold and damp. In Scotland layers of straw or bracken were sometimes packed around the skep and secured by an iron bar across the front. A description of the precautions taken for wintering skeps at West Newton, Tayside, included 'a layer of straw 3 to 4 inches (75-100 mm) thick fastened across the front (of the bee bole) by rope or wire, which enclosed the ruskies (skeps) except at the entrance'. The edges of some boles were rebated to allow boards or doors to be placed across the front to secure the boles for winter. These doors had a flight hole for the bees to enter and leave the skep in warmer weather.

Another method for overwintering hives was to place them in recesses in the cellar of the house or inside an adjacent farm building such as the dairy. This was possible because in cold weather bees cluster together within the hive and do not fly outside it. At Box, Wiltshire, for example, there are five recesses in the garden wall and four larger ones grouped together in one corner of the house cellar. Each of three Tudor manor houses in Kent has a set of gabled bee boles in its garden walls and similarly shaped recesses for winter storage in the cellar. At Henfield, West Sussex, there are eight recesses in a Tudor brick garden wall and in the cellar of the Tudor house are another eight. Four are arched and quite large, being 35 inches (89 cm) high, 24 inches (60 cm) wide and 19 inches (48 cm) deep. The other four are gabled and smaller, similar to the outside boles. It could be that two skeps were housed in each of the larger recesses while the smaller ones were for candles or other lighting. The cellars would be dark and at a fairly constant cool temperature so there was little danger of the bees venturing outside the hives. Keeping the hives in a cool dark place discouraged the bees from consuming their winter honey stores before they could replenish them in the spring.

More often found than boles in cellars,

26

barns, dairies or even farm kitchens are small sheds affixed to garden walls, and detached buildings for the sheltering of skeps during the winter. The latter were for seasonal use and are known as winter bee houses. Most of these are associated with large country houses and date from the seventeenth and eighteenth centuries. They are found in many parts of Britain and eleven had been recorded at the International Bee Research Association by 1983. A substantial stone building 11 feet (3.4 metres) high at its front and set in a wood at Midmar, Inverurie, Grampian, has forty recesses for skeps in the inside walls.

Other buildings could serve the same purpose as a specially constructed winter bee house and in outbuildings such as ice houses skeps would have been placed on the floor or on stands. In the *London Magazine* of 1769, one writer suggested overwintering skeps in an outhouse or ice house: 'Upon their being let out in the warmer air (in spring) they (the bees) recovered immediately and showed in appearance of more strength than the hives did which had been kept out in the usual way!' An outbuilding at Appleby Castle, Cumbria, was traditionally known as 'Lady Clifford's bee house' although there are no recesses inside or outside and no documentary evidence that bees were ever kept there. Perhaps the skeps were placed on the floor as in the ice houses mentioned above.

LEFT: *Recesses for winter storage of skeps in the cellar of a house near Box, Wiltshire.*
TOP RIGHT: *The 'Agricultural Survey for Aberdeenshire' (1811) reported that in order to save on bees' winter provisions it was profitable to keep skeps in an ice house in winter, where the cold would discourage activity and feeding. These recesses in a Wiltshire ice house could have been used for this purpose.*
BOTTOM RIGHT: *St Stephen's, Canterbury, Kent, showing two of nine cellar recesses for overwintering skeps. The house was built in 1490 and has twenty bee boles in a garden wall.*

LEFT: *Under the eaves of an Oxfordshire cottage one of two sets of wooden struts supports a stone skep base and skep. The skep is further protected by a cream pan inverted over the top*
RIGHT: *This sandstone pillar in a Grampian garden could hold one skep in the lower triangular recess.*

# OTHER STRUCTURES

Beekeepers were ingenious in devising ways to house their hives and there are some eccentric as well as extremely attractive examples of this. Some of the bee houses already described are among the latter and perhaps the most intriguing can be seen in a photograph of the late nineteenth century depicting a bee house at Berthddu, Powys, in the shape of a miniature castle. Beside it and in front of a large skep a top-hatted gentleman sits reading in a miniature chair.

Other garden ornamentation included a red sandstone pillar 6 feet (1.8 metres) high and 2 feet (60 cm) across in a Grampian garden. This was built to house a single skep! A more extensive arrangement, perhaps also for bees although there is no direct evidence, can be seen in the grounds of Erddig Hall, Clwyd, in the original plan of 1740. Yew hedges are shown forming eleven adjacent semicircular alcoves. Sited as they are on the edge of the estate and at the furthest distance from the house, these seem more appropriate for the shelter of hives than for the display of statues or other garden ornaments. The hedges have been replanted by the National Trust.

Eighteenth- and nineteenth-century architects such as J. C. Loudon sought to add elegance as well as practicality to their designs for labourers' cottages and an illustration from Loudon's 1840 work, *The Cottager's Garden*, depicts a large attractive gabled alcove with three shelves for skeps. In designing the fantastical set of cottages at Blaise Hamlet, Bristol, John Nash, according to a contemporary, sought to reproduce 'all those little penthouses for beehives, ovens, and irregularities which he found in peasants' cottages and they are so beautiful that it is a sight often visited from Clifton and I have also called it sweet'.

Some skeps were placed inside buildings other than bee houses, for example under window seats in the upper storeys. The bees entered through openings in the outside walls and they then travelled up tubes to concealed hives. At Benthall

28

church in Broseley, Shropshire, the hives were apparently earthenware jars made in a local pottery. The entrance was concealed in the mouth of a carved stone lion head above the door on the south facade of the church. In keeping with the location an appropriate biblical reference (Judges 14: 14) to Samson's riddle was carved above the lion's head: 'Out of the strong came forth sweetness.' A similar system was used at what was formerly (1983) the Nag's Head inn at Avening, Gloucestershire. Four hives within an upstairs bedroom were entered by the bees through holes concealed in a street-facing facade sculpted with garlands. Bees have an age-old connection with brewing, mead being made of fermented honey, and some public houses have used the beehive on their signs since at least 1798.

RIGHT: *'Out of the strong came forth sweetness.' Samson's riddle from Judges 14:14 is illustrated in the south facade of Benthall church, Broseley, Shropshire. The bees entered through the lion's mouth and travelled up a tube to hives within the window seat above.*
BELOW: *An arrangement similar to that at Benthall church. In this former Gloucestershire inn, bees reached their hives in an upstairs bedroom window seat through holes in the carved stone facade.*

# FURTHER READING

There are a number of antiquarian books on beekeeping which include references to bee boles, bee houses, and so on, but most of these are difficult to obtain. The International Bee Research Association (18 North Road, Cardiff CF1 3DY) stocks a wide range of publications on beekeeping and offers some offprints from early works. Additionally, the bulletins of local beekeeping societies and county archaeological journals often contain notes on local sites. The most complete modern treatment of the subject is *The Archaeology of Beekeeping* by Eva Crane (Duckworth, 1983). Other recent publications include:

Crane, E., and Walker, P. 'Evidence on Welsh Beekeeping in the Past', *Folk Life*, volume 23 (1984-5) 21-48.

Desborough, V. F. 'Bee Boles and Beehouses', *Archaeologia Cantiana*, volume 69 (1955) 90-5.

Desborough, V. F. 'Further Bee Boles in Kent', *Archaeologia Cantiana*, volume 70 (1956) 237-40.

Desborough, V. F. 'More Kentish Bee Boles', *Archaeologia Cantiana*, volume 74 (1960) 91-4.

Duruz, R. M., and Crane, E. E. 'English Bee Boles', *Bee World*, volume 34 number 11 (1953) 209-24.

Foster, A. M. 'Bee Boles in Wiltshire', *The Wiltshire Archaeological and Natural History Magazine*, volume 80 (1986) 176-84.

Fraser, H. M. *History of Beekeeping in Britain.* Bee Research Association (London), 1958.

French, K. 'Wide Variety of Bee Boles Can Still be Found in Devon', *Beekeeping, Devon*, volume 26, number 3 (1960) 37-8.

Lambton, L. *The National Trust Book of Beastly Buildings.* Cape, 1985.

Robertson, R., and Gilbert, G. 'Bee Boles' in *Some Aspects of the Domestic Archaeology of Cornwall*, 1979.

Rushen, J. 'Bee Boles and a Beehive's Hum', *Popular Archaeology*, September 1985, 32-8.

Tyson, B. 'Architecture of Lakeland Beekeeping', *Country Life*, volume 171 (1982) 408-9.

Whiston, J. W. 'Beeboles at West Bromwich Manor-house', *Transactions of the Lichfield Archaeological and Historical Society*, volume 4 (1963) 47-51.

Whiston, J. W. 'Bee-boles at Olveston Court, Glos', *Transactions of the Bristol and Gloucester Archaeological Society*, volume 87 (1969) 144-8.

Whiston, J. W. 'Bee-boles at Pipe Ridware Hall Farm, Staffs', *Transactions of the South Staffordshire Archaeological and Historical Society*, volume 13 (1971-2) 43-5.

# PLACES TO VISIT

Most bee boles are on private property and therefore not open to the public. The places listed below are accessible but some do request that visitors telephone beforehand. NT indicates a National Trust property.

ENGLAND

*Appleby Castle Conservation Centre,* Appleby-in-Westmorland, Cumbria. Telephone: Appleby (0930) 51402. (Winter bee house.)

*Attingham Park,* Shrewsbury, Shropshire SY4 4TP (NT). Telephone: Upton Magna (074 377) 203. (Bee house.)

*Benthall Church,* Broseley, Shropshire TF12 5RX. (Beehives once kept in a window seat in a gallery above, with access for bees through the mouth of a stone lion on the south facade of the church.)

*Berkshire College of Agriculture,* Hall Place, Burchett's Green, Maidenhead, Berkshire SL6 6QR. Telephone: Littlewick Green (062 882) 3858. (Victorian bee house.)

*Bretforton Manor,* Bretforton, near Evesham, Worcestershire. Telephone: Evesham (0386) 830216. (Bee house.)

*Cathedral Close,* Canterbury, Kent. (Two groups of bee boles in close.)

*Gainsborough Old Hall,* Parnell Street, Gainsborough, Lincolnshire. Telephone: Gainsborough (0427) 2669. (Bee boles in house wall.)

*Gloucestershire College of Agriculture,* Hartpury House, Hartpury, near Gloucester GL19 3BE. Telephone: Hartpury (045 270) 283 or 284. (Free-standing stone bee shelter.)

*Greys Court,* Rotherfield Greys, Henley-on-Thames, Oxfordshire RG9 4PG (NT). Telephone: Rotherfield Greys (049 17) 529. (Bee boles in garden wall.)

*Lifeboat Public House,* Thornham, Hunstanton, Norfolk. Telephone: Thornham (048 526) 236. (Bee boles in garden.)

*Maidstone Museum and Art Gallery,* St Faith's Street, Maidstone, Kent ME14 1LH. Telephone: Maidstone (0622) 54497. (Bee boles in wall at rear of building.)

*Memorial Gardens,* Westgate, Canterbury, Kent. (Bee boles in garden wall.)

*Milford House Hotel,* Bakewell, Derbyshire. Telephone: Bakewell (062 981) 2130. (Bee boles in house and garden walls. By appointment please.)

*Moseley Old Hall,* Wolverhampton, West Midlands (NT). Telephone: Wolverhampton (0902) 782808.

*National Agricultural Centre,* Stoneleigh, Coventry, West Midlands CV8 2LZ. Telephone: Coventry (0203) 552404. (Headquarters of British Beekeeping Association with display of bee boles and range of wooden hives.)

*Pack of Cards Hotel,* High Street, Combe Martin, Ilfracombe, Devon. Telephone: Combe Martin (027 188) 3327. (Bee boles in garden. By appointment please.)

*Packwood House,* Lapworth, Solihull, West Midlands B94 6AT (NT). Telephone: Lapworth (056 43) 2024. (Bee boles in raised walk at edge of yew gardens.)

*Quebec House,* Westerham, Kent TN16 1TD (NT). Telephone: Westerham (0959) 62206. (Bee boles in garden.)

*Thornbury Castle Hotel,* Thornbury, Bristol BS12 1HH. Telephone: Thornbury (0454) 412647. (Bee boles in walled garden, filled in. By appointment please.)

*Tudor House Museum,* Bugle Street, Southampton, Hampshire. Telephone: Southampton (0703) 224216 or 223855 extension 2768. (Seventeenth-century bee bole and reconstruction of bee shelter in Tudor-style garden.)

*West Woodburn Filling Station,* West Woodburn, Hexham, Northumberland. Telephone: Bellingham (0660) 70241. (Bee boles.)

SCOTLAND

*Arbroath Museum,* Signal Tower, Ladyloan, Arbroath, Angus. Telephone: Arbroath (0241) 75598. (Set of five bee boles moved from original farmstead.)

*Cliffburn Hotel,* Cliffburn Road, Arbroath, Angus. Telephone: Arbroath (0241) 73432. (Bee boles in hotel garden. By appointment please.)

*Garvald Village Hall* (formerly Free Church), Garvald, near Haddington, East Lothian. (Bee boles once in orchard wall, later incorporated in church wall.)

*Pitmedden House,* by Ellon, Aberdeenshire (NT for Scotland). Telephone: Udny (065 13) 2445. (Bee boles in garden wall.)

*St Andrews Cathedral,* St Andrews, Fife. Telephone: St Andrews (0334) 72563. (Bee boles in cathedral ruins.)

*St Salvator's College,* St Andrews University, St Andrews, Fife. Telephone: St Andrews (0334) 76756. (Bee bole in college garden.)

*Tolquon Castle,* Tarves, by Ellon, Aberdeenshire (Ancient Monuments Department). Telephone: Tarves (065 15) 286. (Twelve bee boles in wall in castle grounds.)

*Weaver's Cottage,* Shuttle Street, The Cross, Kilbarchan, Johnstone, Renfrewshire (NT for Scotland). Telephone: Kilbarchan (050 57) 5588. (Bee bole in garden.)

WALES

*Erddig Hall,* near Wrexham, Clwyd LL13 0YT (NT). Telephone: Wrexham (0978) 355314. (Formal gardens with yew hedges thought to have accommodated beehives.)

*Haverfordwest Tourist Information Centre,* 40 High Street, Haverfordwest, Dyfed SA62 6SD. Telephone: Haverfordwest (0437) 3110. (Bee boles in wall behind centre.)

*An arrangement of yew alcoves at Erddig Hall, Clwyd (National Trust), thought to have originally been planted to shelter hives.*

Printed and bound by CPI Group (UK) Ltd, Croydon, CR0 4YY

11/10/2024

01043561-0006